Differentiating Great Lakes Area Native Wild Wolves From Dogs and Wolf-Dog Hybrids

**Differentiating Great Lakes Area Native Wild Wolves
 from Dogs and Wolf-Dog Hybrids**
Copyright 2001 – Beth Duman

Published by Earth Voices Publishing
2512 Sue Drive Howell, MI 48855
EarthVoices.net

Second Edition - 2011

ISBN-10: 0615440479
ISBN-13: 978-0615440477

Cover Photo by Beth Duman – Wolf Hybrid

Differentiating Great Lakes Area Native Wild Wolves From Dogs and Wolf-Dog Hybrids

Beth Duman

Foreword

Wolf populations and distributions have increased drastically across the Great Lakes region of Wisconsin, Michigan and Minnesota in the last 20 years. Although wolves occur mainly in the northern forests of the 3 states, individual wolves have traveled into agricultural areas to the south. Wolves have occurred as far south as Winona in Minnesota and Johnson Creek in Wisconsin. The Johnson Creek area of southern Wisconsin is only 42 miles north of the Illinois border. A small population of wolves also occurs in west-central Wisconsin, and wolves in Michigan may eventually move south into the Lower Peninsula. By the year 2000, over 3000 wolves occurred across the 3 Great Lakes states.

Along with an increase in wolf populations in the Great Lakes states, a concurrent increase has occurred in ownership of wolf-dog hybrids across the region. Every year many wolf-dog hybrids escape into the wild and cause damage, generate concerns, and confuse efforts at determining wolf distribution and abundance. Michigan has recently enacted laws to regulate ownership of wolf-dog hybrids. Wisconsin has proposed regulating wolf-dog hybrids through its wolf management plan, and the Minnesota wolf management prohibits release of wolf-dog hybrids in the wild. (Authors note: As of this 2011 printing, Wisconsin has also adopted policies that regulate the keeping of wolf hybrids)

As a result of the expansion of the wolf population, and increased incidents of problems with wolf-dog hybrids, greater need exists for people to be able to identify these canids. This guide was designed to fulfill that purpose. It is especially designed for wildlife biologists, conservation wardens, wildlife depredation control technicians, animal control officers, and law enforcement officers that occasionally get into situations where they need to be able to positively identify an unknown canid. Because wolves are protected by federal and state regulations, special care needs to be made to properly identify the animal before control actions are taken. Effectiveness of wolf-dog hybrid regulations will depend, in part, on the ability of law enforcement officers to be able to distinguish these hybrids from wolves or dogs.

Beth Duman is well suited for presenting information on the identification of wolves and wolf-dog hybrids. She has been a regional representative for Wolf Park for many years, and has been involved with numerous cases identifying wolves or wolf-dog hybrids in Michigan. Beth was frequently used as an expert witness in cases where injury was caused by wolf-dog hybrids.

Beth has compiled extensive information that she has used for differentiating wolves from dogs and wolf-dog hybrids. She has also consulted and gathered information from other professionals including field biologists, mammalogists, wildlife pathologists and other knowledgeable people. Her work represents perhaps the most comprehensive data set on separations of wolves from dogs and wolf-dog hybrids.

This guide will provide a useful tool to help people differentiate wolves from dogs and wolf-dog hybrid mixes. It provides excellent illustration and narrative to help make decisions on the identity of these canids. Such a tool will be beneficial to wolf conservation and management throughout the Great Lakes region, and other areas.

Adrian P. Wydeven
Wisconsin Wolf Biologist
April 4, 2001

Differentiating Great Lakes Area Native Wild Wolves from Dogs and Wolf-Dog Hybrids

The increase in wild wolf populations has been paralleled by an increase in captive animals that are wolf-dog mixes or are privately kept full wolves. Wildlife biologists are often consulted to identify these animals when the animals are collected after death or occasionally may be asked to verify whether a specific free-ranging animal may perhaps be a wolf or wolf mix. Unfortunately, at this time, there are few decisive DNA tests that are successful at verifying whether a specific animal is a full native wolf. Many clues can be used to discern whether an animal is a wolf or a wolf-dog mix, or if the animal appears to be a full wolf, whether the animal has been captive. Determination of the identity of a specific animal requires careful examination of the animal. If the animal is alive, behavioral clues also need to be considered.

This guide has been assembled to help conservation officers, field biologists, forensic staff, and others learn what are valuable clues in evaluating Canids. This guide alone will not allow determination of whether or not they are native Great Lakes area wolves. Each section discusses a different aspect of native wolves and then lists "clues" that might be identified suggesting that the animal was un-wolf-like in some respect. Obviously, there are many characteristics listed in the descriptions that are not exclusive to wolves. For example, wolves have narrow chests. Many domestic dogs certainly share this trait. If however, a very wolf-like animal was being studied to ascertain whether it was a native wolf or not, a rounded un-wolf-like chest would be a clue that the animal was not a pure wolf but was mixed with domestic dog. Similarly, many dogs have black pads on their feet, but if an animal that was very wolf-like had piebald spotted feet that would be an indication that the animal might not be a full wolf.

This guide is a compilation of observational material that has been reviewed by specialists who work in the field and in laboratories dealing with Great Lakes Area wild wolves both alive and post mortem. It is an attempt to bring together this observational knowledge so other animal specialists can have a framework within which to make educated guesses as to the identity of a specific animal. This guide is not to be used to differentiate dogs from wolf-dog mixes, nor is it intended to differentiate the genetic origin of wild wolves. This guide alone will not allow determination of whether or not they are native Great Lakes Area wolves. The guide simply aids the technician in noticing clues that may signal that a specific animal has characteristics that usually are not present in native wild wolves.

Determination of the identity of an animal requires careful scrutiny.

Morphology and behavior of the entire animal must be taken into consideration. Would it be possible for a native wolf to have uncharacteristic teeth, pelage or body shape? Certainly it would. For a number of years, the highly inbred population of wild wolves in Isle Royale National Park in Michigan contained some individuals that had "thin tails" that never grew the thick "brush" tail characteristic of wolves in the Great Lakes Area in winter. (Peterson 1977) For a number of years, an unusual blue-eyed wolf lived in a wolf pack in Wisconsin. (Ron Schultz pers. comm.) At Wolf Park, a nonprofit educational and research facility in Battle Ground, Indiana, some inbreeding in the long-term captive population has produced individual wolves with shortened, broad-muzzled, neotenous heads and animals with slightly curled tails. As you read through the document, study the corresponding photos and diagrams. Whenever possible, view known specimens of wolves, dogs, and wolf-dog hybrids. Over time, you will become attuned to watching for the clues that will help distinguish wild native wolves.

Body Structure – see Plate 1; Diagrams 1, 2, and 3
A wolf is a tall, slender animal with a narrow keel-shaped chest, proportionally long legs, and large feet. The front legs "toe out" and the rear legs are cow-hocked, the heels of the rear legs point toward each other. A Great Lakes Area wolf can range in size from about 26" at the shoulder to over 30" at the shoulder. The weight varies according to the size and nutritional state of the animal, although a wolf will weigh less than most similar sized dogs because of the proportionally long legs and narrow body. The weight of a Great Lakes Area wolf may range from about 55 lbs if the animal is a small female to occasionally over 100 lbs for a large well-fed adult male. See Plate 1

Clues that the animal may not be a native wolf:
- ❖ The animal has a rounded chest.
- ❖ The animal's legs do not appear proportionally long compared to the body size.
- ❖ The animal weighs proportionally more for its body size.
- ❖ The animal is exceptionally large or small.
- ❖ The animal does not "toe out" in the front or does not have "cow-hocked" rear legs, but has the straighter legged conformation preferred in domestic dogs.

WOLF
(in summer coat)

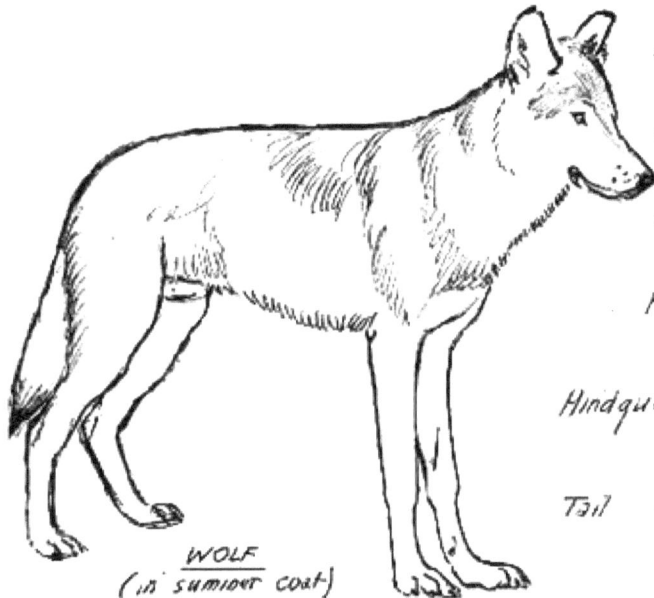

Head : fairly long and narrow.

Muzzle : moderately pointed.

Chest : narrow, lack of depth of ribs.

Feet : Large, long with splayed toes.

Hindquarters : overangulated, cow-hocked.

Tail : carried low, black hair tipped.

MALAMUTE

Head : broad and powerful. Note markings (not present on wolf).

Muzzle : Large and bulky.

Chest : strong, broad and deep.

Feet : compact, toes tight-fitting and well arched.

Hindquarters : powerful. Viewed from behind, the hind leg should not appear bowed nor too close, nor too wide, moves true in line with movement of front line.

Tail : carried over the back.

Diagram 1. Comparison of wolf body shape with that of a malamute dog

9

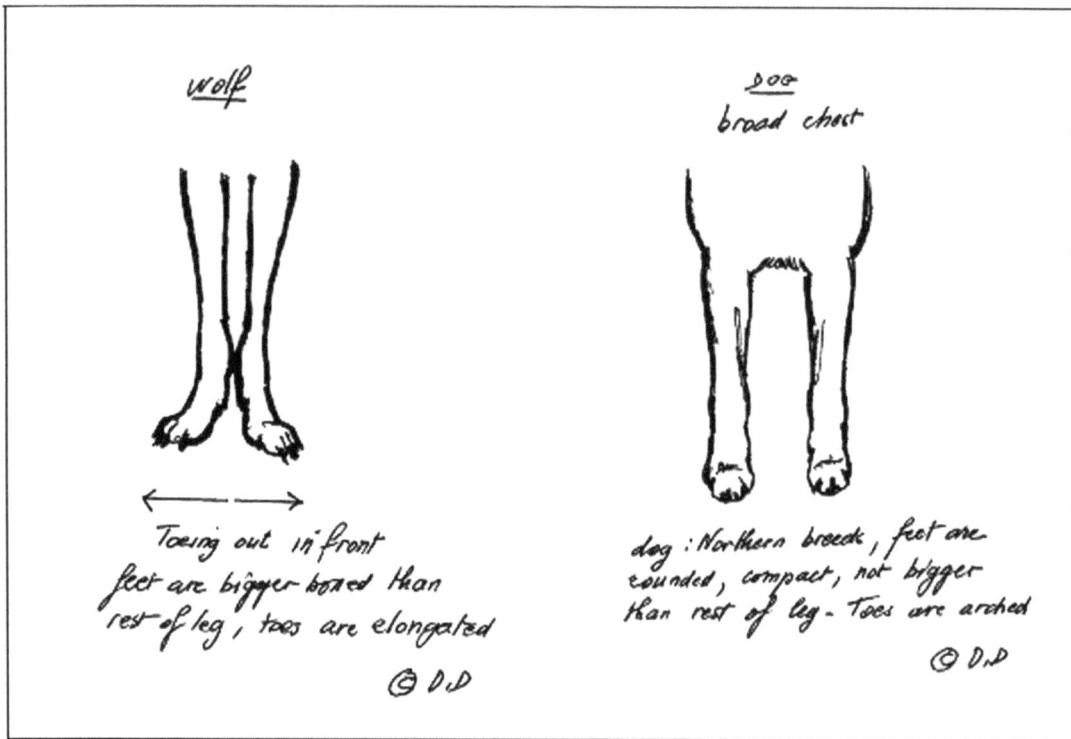

Diagram 2. Seen from the front, wolves "toe out". Some dogs also "toe out" but, in most breeds, a "straight front" is preferred as illustrated.

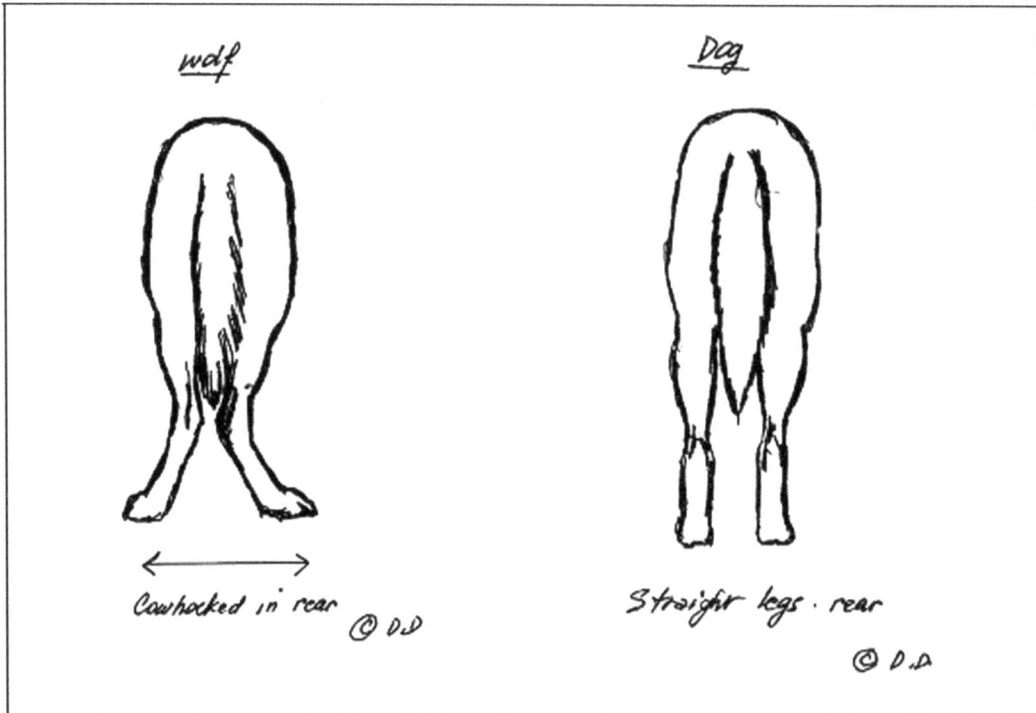

Diagram 3. Seen from the rear, a wolf is "cow hocked"; the heels point in toward each other. Although many dogs are "cow-hocked", most breed characteristics favor a "straight rear".

Plate 1
Body Structure of Wolves

Figure 1. Note the long legs and narrow chest of this captive wolf. The front feet "toe out" rather than pointing straight forward.

Figure 2. Note how the heels point together in this "cow-hocked" faded captive gray wolf.

Figure 3. Siberian Husky -Note the shorter legs and more compact body than that of a wolf.

Figure 4. This wolf-dog mix has shorter legs and a broader chest than a wolf.

11

Head – see Diagram 4; Plate 2

The head of a wolf is large in proportion to the body size. The forehead has a gentle slope without protruding brow, domed forehead or pronounced stop. The muzzle is long, not heavy or broad, but not as narrow as a coyote. The head is wedge shaped but never boxy like many domestic dogs. The jowls are not loose except occasionally in middle age to old animals.

Clues that the animal may not be a native wolf:

- ❖ The head appears domed.
- ❖ The head is boxy shaped.
- ❖ The animal has loose jowls and is not middle aged or old.
- ❖ The forehead has a pronounced stop.
- ❖ The muzzle is broad.

Diagram 4. Comparison of wolf head with malamute head

Plate 2
The head of a wolf compared to the heads of domestic dogs and wolf-dog hybrids

Figure 1. The head of a female
Great Lakes area captive wolf

Figure 2. A Siberian Husky

Figure 3. An Alaskan Malamute

Figure 4. A German Shepherd

Figure 5. Wolf Hybrid
Note the long nose and unusual
ears

Figure 6. Wolf Hybrid
Note the dished face and large
ears.

Eyes – see Plate 3

The eyes of wild wolves in the Great Lakes Area are deep set and are usually yellow or amber in color. Wolf eyes are somewhat slanted and almond shaped. In aged animals, the eyes can appear somewhat more rounded. Wolf pups have bluish eyes that gradually change to the adult color at about four to five months of age.

Clues that the animal may not be a native wolf:

- ❖ The eyes are not yellow or amber in color.
- ❖ The eyes are not deep set
- ❖ The eyes are not slanted.
- ❖ The eye openings including eyelids are rounded in appearance and the animal is not old.

Ears – see Plate 3; Figures 1, 3, 4 and 5

The ears of a wolf are medium sized and erect. The inside of the ears are heavily furred. Grizzled gray wolves that have not severely faded with age will have cinnamon color on the backs of the ears. The tips of the ears are slightly rounded. The cartilage of the ears is thicker than that of dogs. The ears are positioned neither high on the head nor low. Special Consideration: Juvenile wolves "grow into" their ears. Their ears are quite large in proportion to their overall head size.

Clues that an adult the animal may not be a native wolf

- ❖ The ears are overly large.
- ❖ The ears are not heavily furred.
- ❖ The ears are floppy or semi-pricked.
- ❖ The ears are very pointed.
- ❖ The backs of the ears are not cinnamon colored.
- ❖ The ear set is overly high or low.

Nose pad

The nose pad is always black, never pink or pink splotched although recent injury to the nose can cause temporary pink spotting.

Clues that the animal may not be a native wolf:

- ❖ The nose pad is pink.
- ❖ The nose is pink splotched without signs of recent injury.

Whiskers

The whiskers of a wolf of the Great Lakes region are coarse, stiff and black unless the animal has whitened with age. The whiskers closest to the nose will be curled slightly forward. The others will be straight.

Clues that the animal may not be a native wolf:

- ❖ Some or all of the whiskers are white in non-faded animals.
- ❖ The whiskers are not coarse and stiff.
- ❖ The whiskers are all curly.

Plate 3
Wolf eyes and ears compared to those of wolf-dog hybrids

Figure 1. Wolf
Note the angle and color of the eyes of this captive wolf. The ears have rounded tips, are not overly large, and are heavily furred inside.

Figure 2. Young wolf pups have blue eyes that gradually change to the adult yellow or amber eyes by four to five months of age.

Figure 3. This captive wolf pup has not yet grown into his ears.

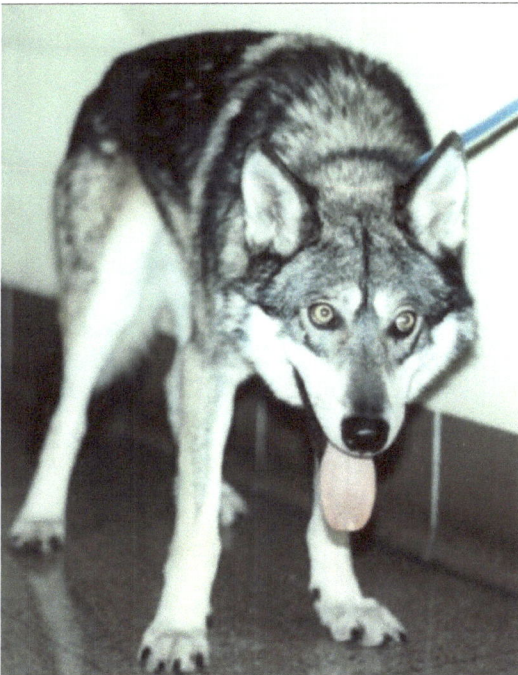

Figure 4. Note the round eyes of this wolf hybrid. The ears are larger than a wolf's and are pointed.

Figure 5. This wolf-shepherd mix had eyes darker than a wolf's. The ears are more heavily furred than a shepherd's but larger than wolf ears.

Dentition and mouth – see Plate 4

In a normal healthy wolf there will be a full complement of 42 teeth. The teeth of wolves are comparatively larger than those of dogs. While dogs will often have tartar on their teeth, the teeth of wolves that feed on carcasses are tartar free. The incisors will usually meet in a scissors bite. The gums near the teeth and the roof of the mouth of wolves have dark pigmentation.

As wolves of the Great Lakes Area are born in late April or early May, the developing teeth of sub adults reflect the age of the pup fairly accurately. The pups should be losing their "puppy" incisors by August. The adult canine teeth should be partially descended by October.

Clues that the animal may not be a native wolf:
- ❖ The teeth are not particularly large and well-formed.
- ❖ The premolars are undersized or some may be missing but not due to wear or injury.
- ❖ There is tartar on the teeth.
- ❖ The gums near the teeth or roof of the mouth are spotted with areas without dark pigmentation.
- ❖ The tongue has black splotches or spots on it. This coloration typifies a Sharpei or chow cross. Some captive wolves at Wolf Park have dark spots on their tongues although this has not been noted in wild wolves.
- ❖ There is an overbite or under bite of the incisors.
- ❖ The change from juvenile to adult dentition does not reflect the time schedule of Great Lakes Area wolf development.

Feet – see Plate 5; Diagrams 5 and 6

The feet of wolves are proportionally very large and long-toed compared to those of domestic dogs. In the winter, hair will protrude between the pads of the foot. Often the hair of the foot will be stained a reddish brown color. The pads of the feet are black. The toenails of Great Lakes Area wolves are black, well formed, and tend to be pointed, especially on the forefeet. There is a significant space between the anterior of the interdigital pad and the posterior of the middle toes, the two middle toes being longer than on most dogs. The anterior portions of the toe pads are less rounded than on most dogs. In adult animals, on the front feet, the foot length, the distance from the back of the interdigital pad to the tip of the longest toe pad, will usually be greater than 8 centimeters. On each fore foot, there will be fifth toe or dewclaw, the "thumb" of the animal. There will be no fifth toe or dewclaw in the rear feet. For information on wolf tracks and tracking see Jim Halfpenny's listing in the bibliography.

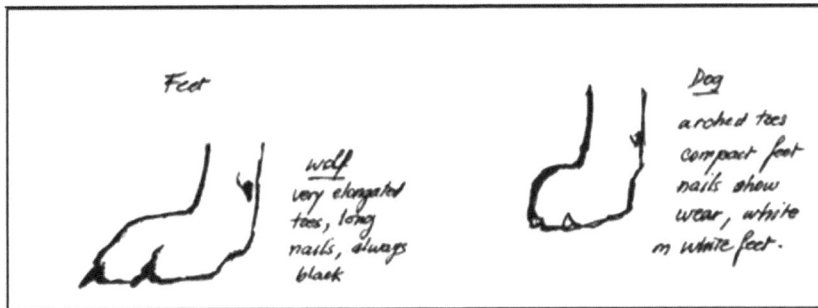

Diagram 5. Wolves have much longer toes than most dogs. Dog breeders prefer a compact foot as illustrated

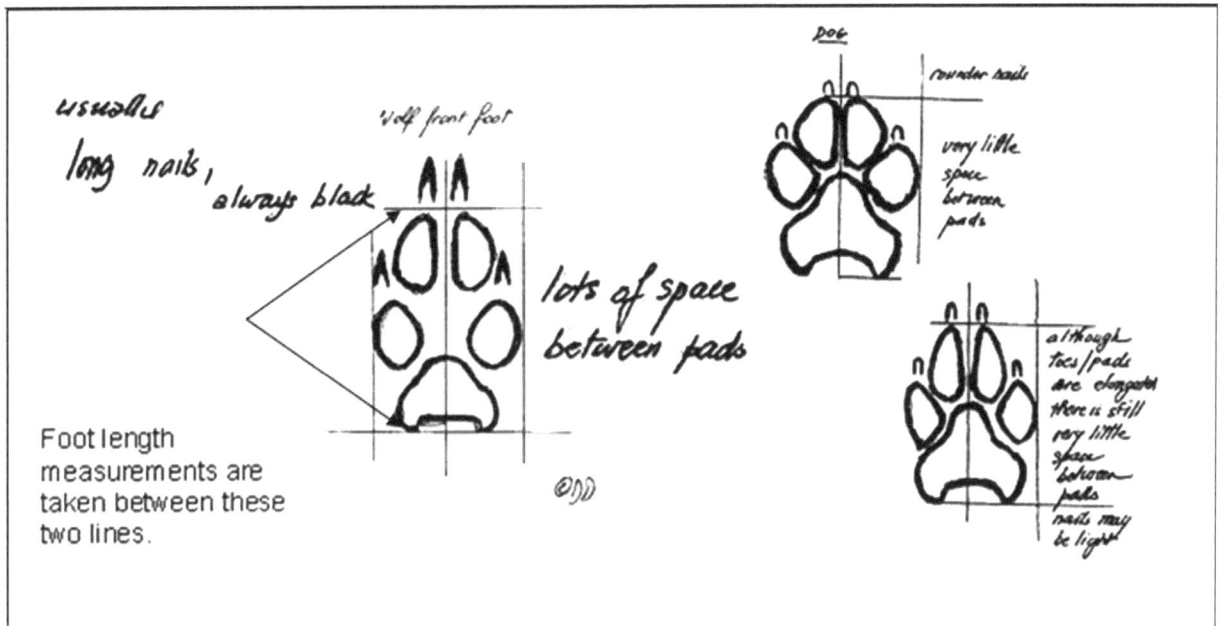

Diagram 6.. The front footpads (not footprints) of a wolf compared with the front footpads of two dogs. Notice that the wolf has toe pads that are not as rounded as some dogs. Also notice the large space between the interdigital pad and the two middle toes.

Clues that the animal may not be a native Great Lakes Area wolf:

- ❖ The feet are not particularly large.
- ❖ The toes are not long.
- ❖ The toe pads are rounded at the front.
- ❖ There is substantial hair growing between the toes in the summer.
- ❖ The pads of the feet are pink or blotched with pink.
- ❖ The toenails are not all black.
- ❖ The toenails are significantly worn, suggesting that the animal has been kept on a hard surface, perhaps in a concrete floored run, or they are clipped.
- ❖ There are no dewclaws on the front feet. Dog breeders often remove the front dewclaws of dogs when they are young pups.
- ❖ There are dewclaws present on the rear feet.

17

Plate 4
Wolf teeth compared with dog teeth

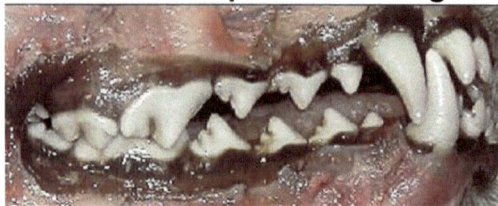

Figure 1. Teeth of an eight-month-old native wolf Notice the well-formed premolars and almost completely descended large canines. The gums near the teeth have dark pigmentation. (This photo is of a road-killed partially necropsied specimen)

Figure 2. Teeth of a wolf hybrid
This also is a necropsy specimen that was partially identified as a non-native wolf because of its small, widely spaced teeth.

Figure 3. Teeth of an eight month-old native wolf
Notice the dark pigmentation of the roof of the mouth and gums near the teeth.

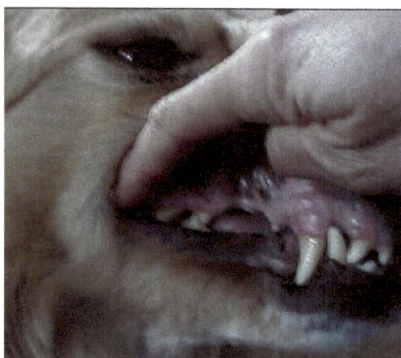

Figure 4. Teeth of a mixed breed dog Note the small widely spaced teeth and missing premolar

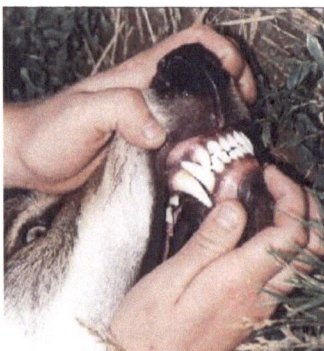

Figure 5. Teeth of an adult captive wolf Notice the black nose and the lack of spotting and dark pigmented gums near the teeth

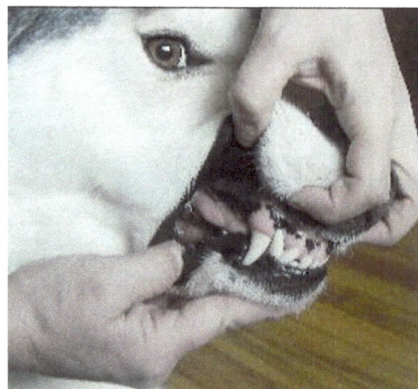

Figure 6. Note the piebald spotted gums of this Alaskan Malamute.

Plate 5
Wolf feet compared with dog and wolf hybrid feet

Figure 1. A captive wolf
Notice the very large feet with black toenails.

Figure 2. Front foot of a native wolf in winter. Note the space between the interdigital pad and the long middle toes.

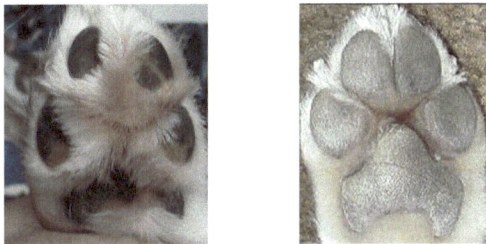

Figure 3. Very un-wolf-like dog feet
Note the rounded toe pads. The malamute foot on the left also has piebald spotting.

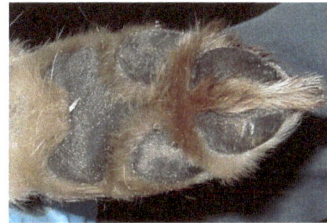

Figure 4. Rear foot of native wolf in winter
Note the pads are all black.

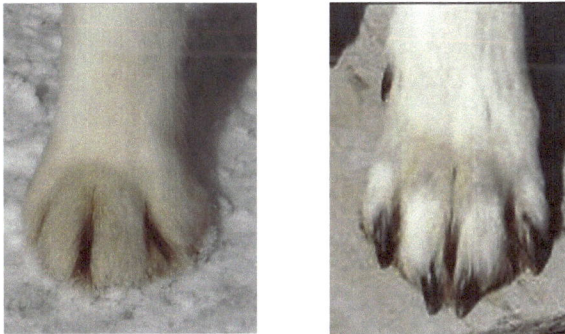

Figure 5. German shepherd front foot
This particular dog has toe pads that are more pointed.

Figure 6. Siberian husky front foot on left compared with wolf front on right
This dog foot is proportionally much smaller than the wolf foot and has white toenails. Note that there is no "thumb" or dewclaw on the dog. Many dog breeders remove the dewclaws when the dogs are young pups. The wolf foot has long toes and black toenails. The front foot of a wolf will have a "thumb" or dewclaw. Although dogs often have rear dewclaws also, these have not been seen native wild wolves.

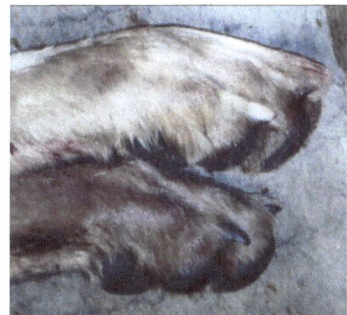

Figure 7. Wolf hybrid rear feet
This wolf-like animal had midsize feet and some white toenails.

Tail – see Plate 6; Diagram 7

The tail of a wolf is a "brush tail' tapering slightly at the end. Fall through spring, the tail will be heavily furred. In late spring through summer, depending on the shedding pattern of the individual animal, it may be thinner with short hair. On the top of the tail, a few inches down from its base, there is a patch of thick wiry black hair marking the location of the precaudal gland. The tail will not be curled and will normally be held flaccid unless the animal is displaying or moving quickly.

Clues that the animal may not be a native wolf:
- ❖ The tail is not "brush" shaped.
- ❖ The tail is "flag" shaped.
- ❖ The tail is curled.

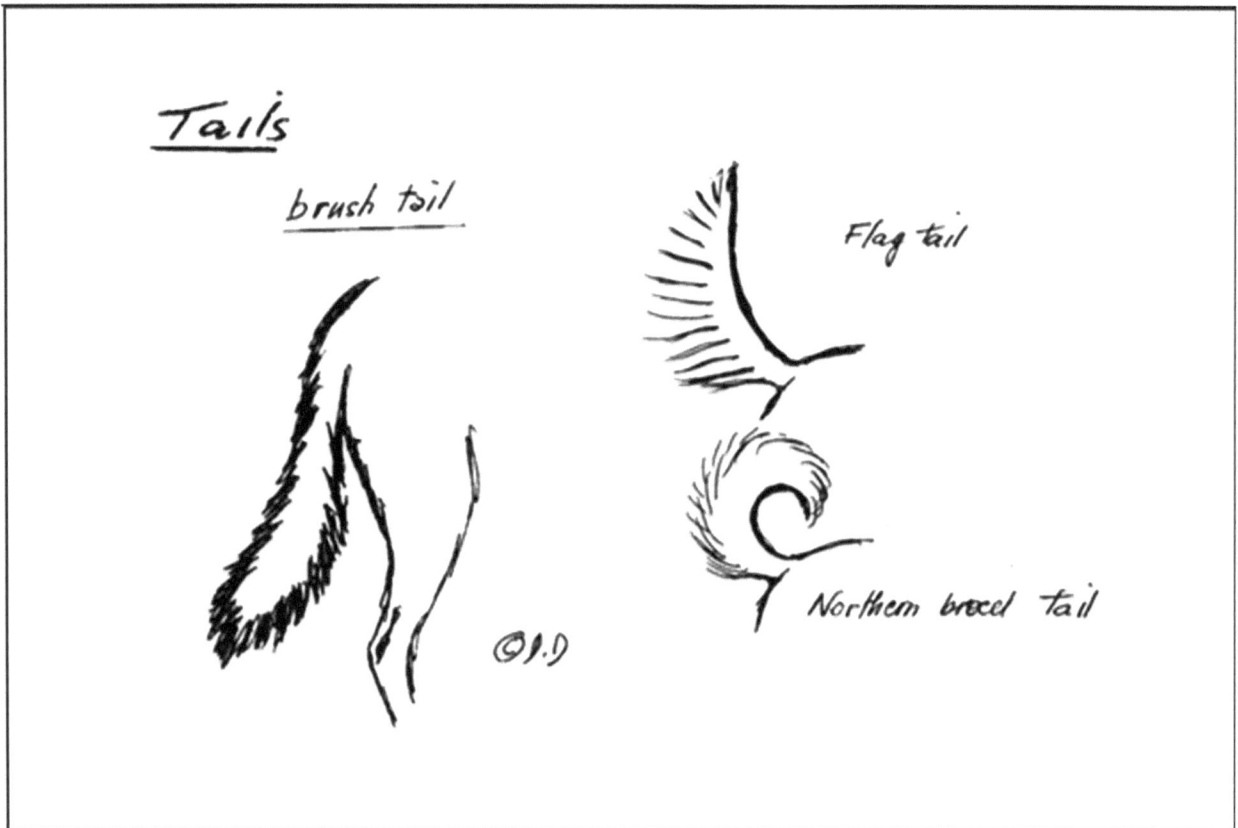

Diagram 7. Wolves will have a "brush tail" that does not curve over the back. Many dogs have "flag tails". Northern breed dogs such as Alaskan malamutes and Siberian huskies have more wolf like "brush tails" but usually carry them curled over their backs when they are active.

Plate 6
Wolf tails compared with dog and wolf hybrid tails

Figure 1. Captive wolf
Notice the black-tipped brush tail.

Figure 2. When interacting socially, wolves may hold their tails higher but the tails will never be curled as they are in many dogs.

Figure 3. Siberian huskies
Note the curly tails.

Figure 4. This brown Belgian sheepdog has a flag tail

Figure 5. This Wolf-Great Pyrenees mix has a very wolf-like tail

Figure 6 This wolf hybrid has a curly tail

Coat Color – see Plates 7 and 8

Great Lakes Area native wolves exist in two basic color phases; grizzled gray and black. Wolves start out darker in coloration and lighten in color as they age, some lightening more than others.

Grizzled gray wolves begin life as sooty dark pups. By the end of the summer, their coats will have lightened to the typical "gray" color, but they can retain black coloration on their chins and darker tones on their front legs. Adult Great Lakes Area gray wolves will have grizzled gray coloration dorsally, lightening ventrally to light tan or light buff on their under-throat and belly. The end of the tail will be black. Their faces will have darker masks that are not as distinct as most on some domestic breeds of dogs. The facial mask extends gently down between and below the eyes fading into light brown tones on the bridge of the nose. There is a lighter patch above each eye. In any given animal, the coat can change considerably in color from grizzled gray in the winter to brownish in the summer. The photos of captive wolves in this document are animals from Wolf Park in Battle Ground, Indiana. Great Lakes Area native wolves usually have more light brown tones in their coats and black shading down the fronts of their forelegs than do some of the Wolf Park animals, although native wolves can lighten or "fade" as they age.

Black wolves will start their lives as black pups with a white blaze on their chests. As black wolves age, they lighten to various degrees. None remain completely black throughout their lives. Most often, the animals first show whitening of their feet and muzzle areas but usually retain the facial mask. Some black wolves in captivity have become almost white in their latter years. Black wolves are generally rare in the Great Lakes region, probably representing less than 5% of the population, but in some packs more than half of the members are the black color phase. In late spring and summer, black wolves' coats will often show bleached maroon color hair.

Clues that an animal may not be a native wolf:
- ❖ Northern dogs usually are born with lighter colored pups that darken with age. Any juvenile that is light in color would not be a native wolf.
- ❖ The animal is uniformly black or white.
- ❖ The undercoat is white or brown.
- ❖ Exaggerated white patch above each eye
- ❖ The facial mask has defined edges without gradual brown shading on the nose.
- ❖ The backs of the ears are not cinnamon colored.
- ❖ The animal is grizzled gray with white hairs at the end of the tail. Captive older faded black wolves often have white on the ends of their tails although it has not been reported in wild specimens.
- ❖ The animal has "freckles" on legs or face.
- ❖ The under belly, legs, or cheeks are pure white rather than light tan unless the animal is obviously an aged-faded individual.

Plate 7
Wolf color- Grizzled gray wolves compared with wolf hybrids

Figure 1. A captive grizzled gray wolf in winter

Figure 2. A wild Michigan wolf in summer Note the brown coloration.

Figures 3. Grizzled gray wolves begin their lives as dark puppies and lighten as they get older.

Figures 4 & 5. Some grizzled gray wolves lighten or "fade" considerably as they age. These almost white captive wolves were grizzled gray when they were younger.

Figures 6 – 8. Wolf hybrids
Wolf hybrids can often have unwolflike coloration, especially in the facial mask. Quite often, the facial mask will be darker or better defined than a wolf's facial mask, or lack the brown shading on the nose.

Plate 8
Black color phase wolves

Figures 1 – 6. All of these captive animals are black phase wolves that started out as black pups like the animal in the upper left corner. In the bottom left photo, see a pair of "blue wolves", partially faded black wolves. Note the more round appearing eyes of the whitened old wolf in the photo on the bottom right.

Pelage – see Plates 9 and 10

The coat of a wolf consists of two layers, the coarse and longer guard hairs and a very soft, dense, wooly undercoat. The coat will vary seasonally. The late fall to winter coat will be very dense. The undercoat will usually be gray, although it may be lighter gray on older animals. The guard hairs on the head and trunk of the animals are banded. The thick guard hairs of the hackles (the longer hairs over the shoulders) form a "cape" over the shoulders of the wolf and are usually 4-banded on grizzled gray-colored or faded black wolves: black tip -- light band – dark band – grayish at the base.

Beginning in late spring and moving into summer, the wolf will lose its dense undercoat in thick clumps, the shedding usually starting on the legs and belly. By mid summer, an adult wolf will no longer have any undercoat and the guard hairs will lie flatter on the body. During the summer months, the guard hairs will also shed as the new coat begins to grow. By mid to late autumn, the animal again will be in full coat.

Special Consideration: The first winter coat of a wolf may look scruffier than the coat of an adult wolf.

Clues that the animal may not be a native wolf:

❖ The undercoat is not gray or light gray.
❖ The animal is shedding at the wrong time of the year or has significant undercoat in summer months.
❖ The guard hairs on the body are not banded. Many domestic dogs, especially the majority of Siberian huskies and Malamutes will have dark tipped white guard hairs. Most of them will not have 4 color bands on the hackle hairs.
❖ The hair is excessively long or short.
❖ The hair is wavy or curly.
❖ The winter undercoat is not extremely dense.
❖ There are broken guard hairs in the neck area indicating that the animal has worn a collar. (Unless, of course, it had been radio-collared for study)

Reproductive System

Wolves are seasonal breeders that usually do not reproduce until their second year. Adult breeding females will be in season in January through February. Estrus blood of female wolves occurs up to 45 days before estrus and can be present from late December to early March. Non-breeding adult females will not come into season. The pups will be born in the spring, usually in late April or early May. A female will be lactating from the birth date of pups possibly into early July. When examining an animal to determine whether it is a native wolf, this breeding cycle must be taken into consideration. In male wolves, the testicles are fully haired and carried close to the body. Male wolves have little or no viable sperm from April until late October, and the testes will be relatively small during this period.

Clues that an animal may not be a native wolf:

❖ A female is in season, is pregnant, or is lactating during the wrong time of the year.
❖ The testicles of a male are not fully haired or carried close to the body.
❖ The animal has been castrated or spayed.

Plate 9
Wolf hair compared with dog hair

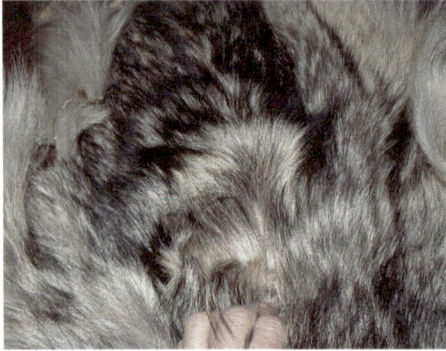

Figure 1. Hackles of a grizzled gray Great Lakes Area wolf
Note the four color bands: black tip – light band – dark band – grayish at the base.

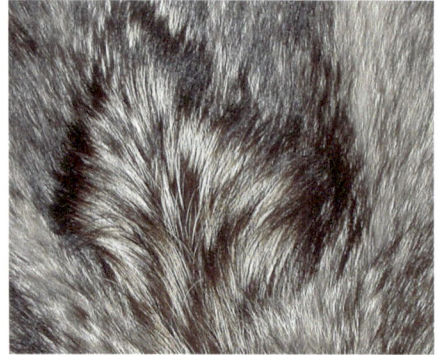

Figure 2. Hackles of a "faded" black wolf (from an Alaskan wolf pelt)
Note that the individual hairs have four bands like those of the grizzled gray

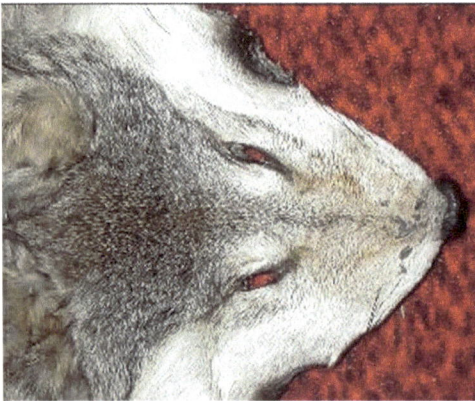

Figure 3. Great Lakes Area Wolf pelt showing gentle shading and brown tones of the facial mask

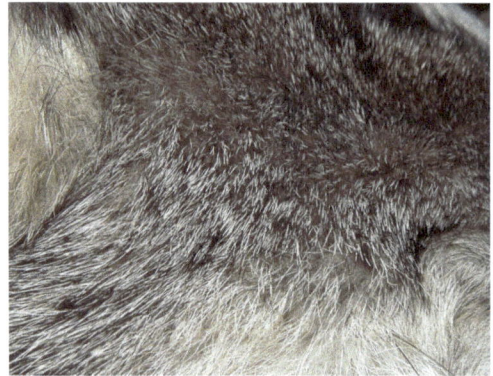

Figure 4. Close-up view of the head of the same wolf pelt
The guard hairs on the head are also four-banded

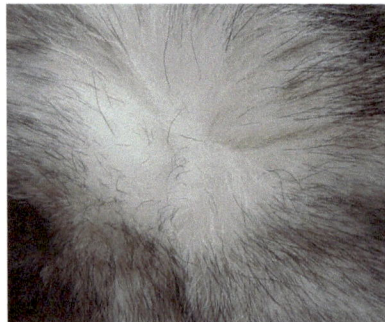

Figure 5. Close-up view of Siberian husky hair
The most common banding pattern for Siberian huskies and Alaskan malamutes is two bands: dark tip – white.

Plate 10
Seasonal variations in wolf pelage

Figure 1. Captive wolf in winter

Figure 2. Wild Michigan wolf in summer

Figure 3. The hair of this faded black phase captive wolf is shedding out in thick clumps.

Figure 4. Wolves have long hair in the winter and short hair in the summer. The profuse shedding that begins in late spring starts on the belly and legs. This is an older faded grizzled gray captive wolf.

Figure 5. Fully shed out wolves are often mistaken for deer in the wild because of their extremely long legs and thin bodies.

Figure 6. In the winter, wolves look heavier and shorter legged because of their longer coats.

Stomach Contents

When doing necropsies on wolf-like animals, the contents of the stomach need to be checked. Native wolves will most likely have deer tissues, hair and bones in their stomachs, although other game and wild fruits might also be present.

Clues that the animal may not be a native wolf:
❖ The animal has no food in its stomach or digestive tract indicating that the animal might have been dumped or strayed and might be unable to find food.
❖ The animal's stomach contains dog food or other prepared food.

Skull – see Plate 11, Diagrams 8 and 9

The cleaned skull of an animal can be examined to give clues as to the identification of it as a wolf, dog or wolf mix. If the whole body has been examined prior to cleaning the skull, sufficient clues have most likely already been recognized identifying the animal. Looking at the skull, in that case, further supports the conclusions already drawn in examining the body. In some cases, however, if only the skull is available or the animal is particularly difficult to identify, examination of the skull can be critical in identifying the animal. This should be done by a trained osteologist.

Linear discriminate analysis, a statistical method that compares standard skull measurements, has been used successfully to identify wolves, wolf mixes and dogs. The method is complex and is best described by Walker and Friesen (1982) and Lawrence and Bossert.(1967). Some skull measurements may be possible with x-rays of live wolves. (Walker pers. comm.)

Gross examination of the skull can give other clues that can help in identification. The overall shape of the skull of wolves and most dogs is quite different. If an animal is part dog, the skull often will reflect the difference in shape. Most often, an animal mixed with dog will show more of a furrow between the eyes and a dishing of the face. The skull length of a wolf is generally 23-27 cm (9.1 to 10.6 inches). If the skull length is less than 23 cm or 9 inches, the animal is probably a coyote or a dog or dog mix.

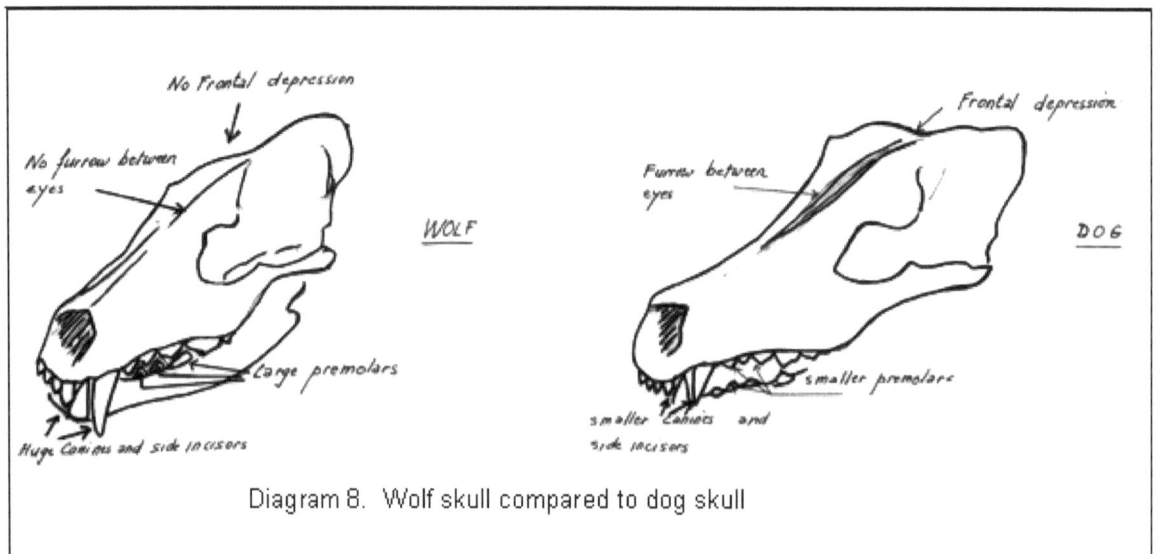

Diagram 8. Wolf skull compared to dog skull

N. A. Ilgin (1941) cited that the orbital angle of an adult wolf differs from that of a dog. The orbital angle is the measured angle between a plane (straight line) drawn across the top of the skull and a plane (straight line) drawn from the top of the skull out across the zygomatic arch. The orbital angle of wolves characteristically is 40° - 45° and the orbital angle of most dogs ranges from 53° - 60°. A wolf dog mix would be expected to fall between these ranges.

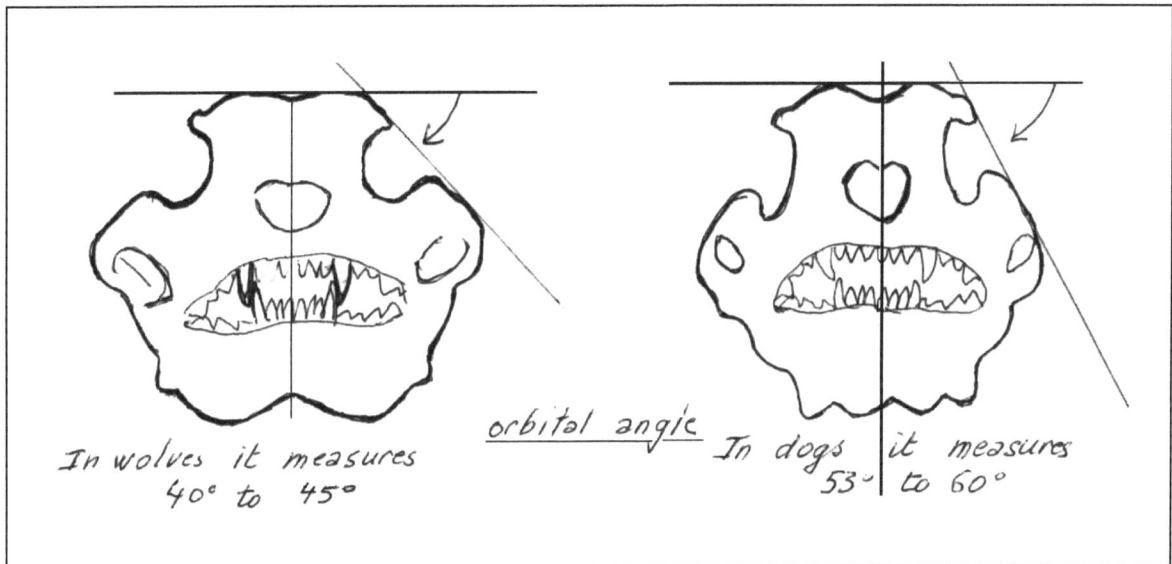

In wolves it measures 40° to 45°

orbital angle In dogs it measures 53° to 60°

Diagram 9. The Orbital angle of a wolf skull compared to a dog skull

Another characteristic that differs between wolf and dog skulls is the shape of the auditory or tympanic bullae. In wolves, these structures on the bottom of the skull that contain the inner ears of the wolf, are rounded. In most dogs they are flattened or less fully formed. A wolf mix will usually show some flattening of the auditory bullae.

Blood Titers
When an animal has been exposed to and has survived a disease, antibodies will remain in the animal's body. When an animal is vaccinated against disease, antibodies will also be produced. Blood titer tests can disclose what diseases an animal has been exposed to and what diseases it has been vaccinated against. Veterinary laboratories can run titer tests for some of the common diseases for which domestic dogs are vaccinated. Great Lakes Area wolves might have measurable titer levels for some common dog diseases such as parvovirus, leptospirosis and distemper. About 1/3 of wild wolves tested in Wisconsin show positive titer levels for canine parvovirus and about 1/80 show titer levels for exposure to distemper. Statistically speaking, the chance for a wild wolf to show titer levels for exposure to both distemper and parvovirus would be 1 in 240 animals tested. If the animal had been tagged or radio collared by wildlife biologists, it would be necessary to check capture records to see if the animal had been vaccinated for distemper and parvovirus at that time

Since rabid wolves would die of rabies rather than producing antibodies and surviving, and wildlife biologists are not vaccinating Great Lakes Area wolves for rabies at this time, a positive rabies titer would indicate that the animal was a captive that had been vaccinated.

Plate 11
Wolf skull compared to dog and wolf hybrid skulls

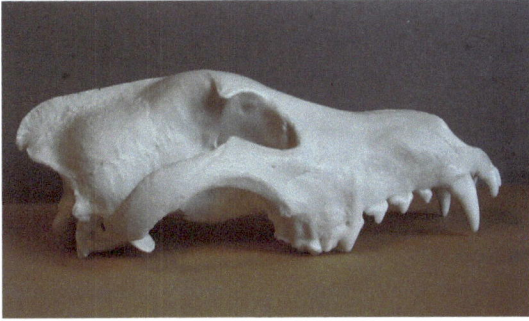

Figure 1. A wolf skull (actually a museum replica)

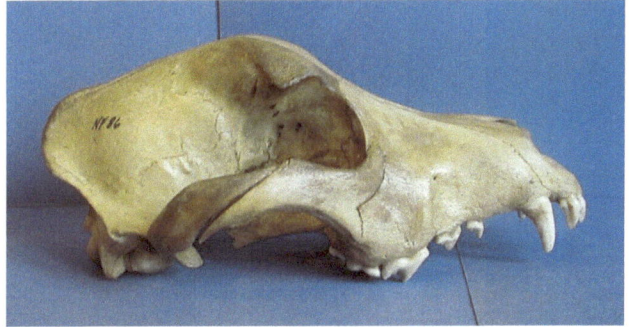

Figure 2. A Newfoundland skull
Note the high domed head of this specimen.

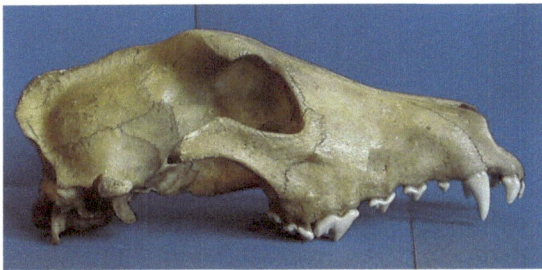

Figure 3. A collie-shepherd mix skull
Note the more widely spaced premolars than the wolf skull.

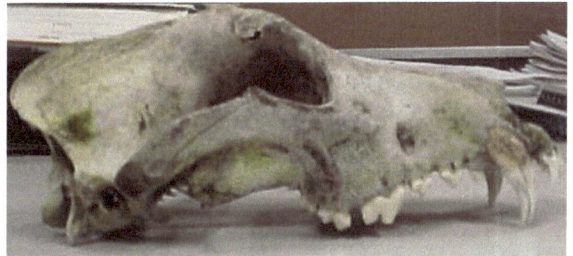

Figure 4. Wolf hybrid skull
Notice the very large canine teeth similar to a wolf but the domed skull more like the Newfoundland dog

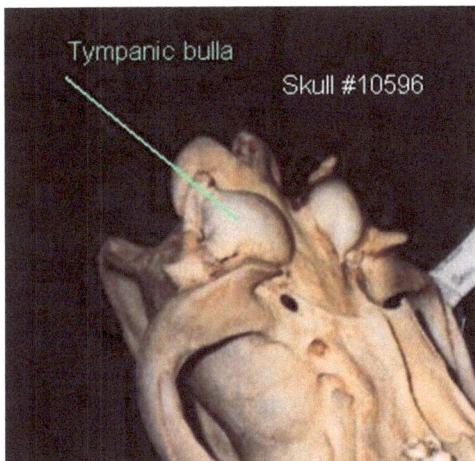

Figure 5. A wolf skull with auditory or tympanic bulla labeled
Wolves have rounded tympanic bullae.

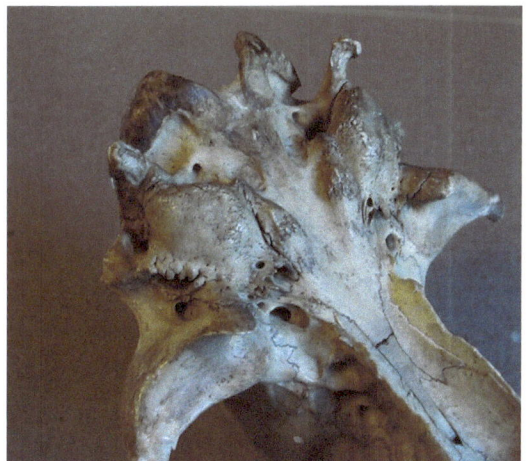

Figure 6. The tympanic bullae of dogs are more flattened than those of wolves as seen in this photo of a collie shepherd skull.

31

Clues that the animal may not be a native wolf:
- ❖ The animal shows titer levels for both parvovirus, and distemper exposure and is not an animal that has been captured, tagged and studied.
- ❖ The animal shows titer levels for rabies exposure.

Tattoos and Microchips

If possible, all suspect animals should be checked for tattoos and microchips. Tattoos usually would appear on the inguinal area of the stomach, the inner thigh, or in the ears. Microchips normally would be found on the dorsal area between the shoulder blades. In Michigan, all hybrids being kept legally will be micro-chipped with an ID number that can be traced to the owner.

Clues that the animal may not be a native wolf:
- ❖ The animal is tattooed.
- ❖ The animal is identified with a microchip number traceable to a private individual rather than a state wildlife agency.

Behavioral Clues

Often the circumstances surrounding where the animal was sighted or has been found dead give clues to whether the animal was a native wild wolf or not. Although there have been some "fearless" wild wolves, in general native wolves will not be comfortable around humans and their dwellings. Wild wolves usually interact aggressively with domestic dogs. Wild wolves have killed domestic dogs and have preyed on farm animals. Using clues to assess whether a specific animal is a native wolf will often require interviewing residents of an area where the animal has been seen. Often, neighbors in an area will know of someone who has been keeping captive wolves or wolf-dogs. They may also have had experiences with the animal that would give clues as to the animal's origin. When wild wolves make kills they are efficient, usually killing the animal with bites to the throat and/or hindquarters. Animals that have been in captivity will be less efficient, usually killing the animal with multiple bites all over the body.

Clues that an animal may not be a native wolf:
- ❖ The animal is consistently seen hanging around human populated areas.
- ❖ The animal interacts playfully with domestic dogs.
- ❖ The animal approaches humans and attempts to interact with them socially.
- ❖ The animal is in an area where hybrid breeders or owners are known to live and occasionally have animals escape.
- ❖ Livestock has been killed inefficiently with multiple bites all over the body of the victim or the animal has been killed and not eaten.
- ❖ The animal kills livestock without eating them.
- ❖ The animal is easily trapped in a cage type trap.

Vocalization
Wolves are known for their howls, but can also growl, whimper whine, and bark. Unlike most dogs, however, wolves do not bark a lot. The barks of wolves are a breathier "hoof hoof" sound. When upset, they may also make a bark-howl combination.

Clues that the animal may not be a native wolf:
❖ The animal's bark is a repetitive doglike "woof woof".
❖ The animal makes friendly greeting barks
❖ The animal does "recreational barking".

Conclusion
Each animal needs to be examined as closely as possible, taking all of the aforementioned information into consideration. An animal that is not a native wolf will probably give the examiner a number of clues that will aid in making an educated evaluation of the animal. Take time to go over each animal carefully. Over time, the clues will become more obvious as you have studied a number of animals. Listed in the "resource" section of this paper are a number of zoos and nature centers in the Great Lakes Area that have captive wolves and wolf hybrids. Observe as many animals as you can but be aware that many animals in captivity, whether through inbreeding or possible hybridization with domestic dogs, do not give accurate representation of our native wolves.

Places to View Captive Wolves and Wolf Hybrids in the Great Lakes Area:
Note: Some animals displayed as wolves at some of these facilities do not fully resemble wild wolves and may have color, color pattern intensity, or body shape variations differing from wild wolves because the animals come from inbred captive stock or from bloodlines of captive animals that may contain possible small amounts of dog ancestry. When in doubt, refer to the photos in this publication as they have been carefully chosen to depict the characteristics of native Great Lakes Area wild wolves.

Michigan
Binder Park Zoo, Battle Creek, MI – Mexican Wolves
Saginaw Zoo, Saginaw, MI – Wolves
Garlynn Zoological Park, Naubinway, MI – Wolf Hybrids
Potter Park Zoo, Lansing, MI - Wolves

Minnesota
Wildlife Science Center, Forest Lake, MN – Wolves & Wolf Hybrids
International Wolf Center, Ely MN – Wolves
The Minnesota Zoo, Apple Valley, MN –Mexican Wolves & Gray Wolves
Como Park Zoo, St. Paul, MN - Wolves

Wisconsin

Mackenzie Environmental Center, Sun Prairie, WI – A Wolf & a Wolf Hybrid
Milwaukee County Zoo, Milwaukee, WI – Great Lakes Area Wolves
Northeast Wisconsin Zoo, Suamico, WI – Red Wolves
Lincoln Park Zoo, Manitowoch, WI - Wolves

Plate 12
Captive wolves differing in appearance from wild wolves

Figure 1. Captive bred wolves can show atypical body shape and color patterns perhaps because of inbreeding or possible hybridization with domestic dogs. Note the unusual shortened muzzle of this animal.

Figure 2. This captive bred animal shows a neotenous or puppy-like head with a shortened muzzle and underbite.

Figure 3. Compare the typical looking captive bred wolf on the left in this photo with the atypical animal on the right. Note the more intense facial markings and broader body.

Plate 13
Wolf-dog hybrids

Figure 1. This animal resulted from crossing a dark colored malamute with a full wolf.

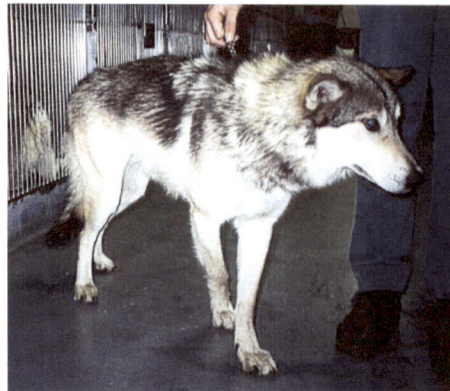

Figure 2. A wolf/husky mix

Figure 3. High % hybrid

Figure 4. A generic hybrid

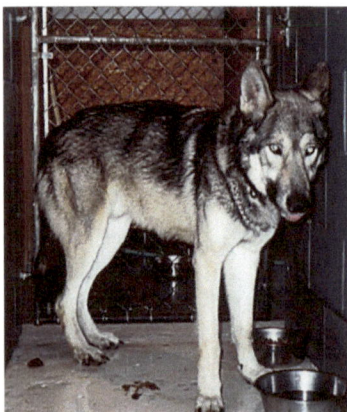

Figure 5. Small feet, and doglike coat, are clues to this hybrid's identity

Figure 6. A wolf/elkhound mix

Credits

Thanks to all who reviewed this document offering suggestions, and help:
Bob Duman, Kate Runyon, Monty Sloan, Adrian Wydeven, Ron Schultz, Nancy Thomas, John Weller, Peggy Callahan, Tom Cooley, Jim Hammill, Pam Troxell, Pat Goodmann, Brenda Aloff, Monty Sloan, Bonnie Yates, Jan Koler, Cindy and Dan O'Malley, Michele Thie, and Danny Walker.

Thanks to all who allowed the use of their photos and art work
Jill Moore – wolf sketch – title pages
Monty Sloan and Wolf Park – all of the photos of captive wolves except for Plate 2, figure 1; page 48 – figures 1, 2, 4, & 5
Indra Sanford – Plate 1, figure 4; Plate 7, figures 5 & 6; Plate 14, figure 3
Adrian Wydeven – Plate 11, figure 4, page 43 - figure 1
Jim Hammill –Plate 7, figure 2; Plate 10, Page 48 – figures 3 & 6
All other photos were taken by the author.

Thanks to artist Daniele Daugherty for the original line drawings.

Thanks to those who offered their time and dogs for photography:
Michele Thie and Cindy and Dan O'Malley

Bibliography & Resources

Duman, Elisabeth. 1993. Is it a Wolf and What Will It Do? 1993 Michigan Veterinary Conference Proceedings, DMS/LAS pp.1-4.

Duman, Elisabeth. 1994. A Wolf At Your Door? Michigan Humane Society, Volume 20, Number 3, pp 1 & 9.

Halfpenny, Jim. 1998. Tracking Canids: Track and Trail Analysis, A Naturalists World, P.O.
Box 989, Gardener, MT 59030.

Halfpenny, Jim. 1995. Tracking Wolves: Basics (Slide show or Computer Disk), A Naturalists World, P.O. Box 989, Gardener, MT 59030.

Iljin, N.A.1941. Wolf Dog Genetics, Genetics, Vol.42, No. 3, pp.359-414.

Lawrence, Barbara and William H. Bossert. 1967. Multiple Character Analysis of Canis lupus, latrans, and familiaris , With a Discussion if the Relationships of Canis niger. American Zoologist, 7:223-232.

Lawrence, Barbara and William H. Bossert. 1969. The cranial evidence of hybridization in New England Canis, Brevoria 330:1-13.

Lawrence, Barbara and William H. Bossert 1975. Relationships of North American Canis shown by a multiple character analysis of selected populations. In (M. W. Fox, ed.) The Wild Canids, New York: Van Nostrand-Reinhold, pp. 73-86.

Moore, Tom D., Danny R. Walker and Rachel A. Fisher, PhD. Wolves, Dogs and Hybrids: Law Enforcement Problems in Wyoming, Western Proceedings 63:69-81.

Peterson, Rolf Olin. 1977. Wolf Ecology and Prey Relationships on Isle Royale, National Park Service Scientific Monogram Series, Number Eleven.

Tucker, P.A., D.L. Davis, and R.R. Ream. 1999. Wolves:Indentification, Documentation, Population Monitoring and Conservation Considerations. Northern Rockies Natural Resource Center, National Wildlife Federation, Missoula, Montana

Von den Driesch, Angela.1976. A Guide to the Measurement of Animal Bones from Archeological Sites, Peabody Museum Bulletin 1, Peabody Museum of Archeology and Ethnology, Harvard University.

Walker, Danny and George Frison.1982. Studies on Amerindian Dogs, 3: Prehistoric Wolf/Dog Hybrids from the Northwestern Plains, Journal of Archeological Science1982, 9, 125-172.

Plate 14
What clues can you find that tell you that these wolf-like animals are part dog?
See page 40 for a description of each animal.

Figure 1.

Figure 2.

Figure 3.

Figure 4

Figure 5.

Figure 6.

Discussion of animals in Plate 14

Figure 1: This animal is an elkhound/wolf mix. Notice the broader chest and unwolflike dark mask.

Figure 2: The coat of this animal is doglike. The head is more domed with a pronounced stop.

Figure 3: Note the unusual ears of this animal, the lack of facial mask, and the doglike coat.

Figure 4: The up-turned tail and coat of this animal show dog background in this animal. Note also the pointed ears.

Figure 5: Notice the black and white coloration with lack of brown tones in this animal. The facial mask is overly defined without gradual shading. Note also the broad chest.

Figure 6: Note the broader chest and unusual facial markings of this animal. The muzzle is broader than a wolf's.

Canine Identification Check-Off

Date_____ Species_____

Date:_____ Breeder:_____

Owner:_____ Age:_____

Comments:

Checklist of wolf-like characteristics:
- o Narrow keeled chest
- o Proportionally long legs
- o Weight between 50 and 110 pounds
- o Front legs "toe out"
- o Long muzzle
- o Lack of facial stop
- o Eyes yellow or amber
- o Eyes are slanted
- o Medium sized ears
- o Ears heavily furred
- o Rounded ears, erect
- o Cinnamon color on back of ears if a grizzled gray color wolf
- o Large, well-formed teeth
- o No tartar on teeth
- o Gums dark pigmented near the teeth
- o No black splotches or spots on tongue
- o Roof of mouth is evenly dark pigmented
- o Black nose pad
- o Robust, straight whiskers on cheeks
- o Foot pads at least 8.0 cm long (not measuring claws)
- o Pads of feet black
- o Toenails black
- o Brush tail
- o Black tip on tail
- o Black spot (precaudal gland) on back of tail
- o Banded hair over body
- o Hackles hair 4-banded
- o Dense gray undercoat if winter animal
- o Animal is not uniform black or white
- o Facial mask with gradual brown shading on nose (except in black color phase wolves)
- o Underbelly gradually goes from tan to cream color, not pure white

How to take Wolf Measurements

Nose to Tip of Tail: Run a tape the length of the body stopping with the end of the bone in the tail. Take all measurements in centimeters or inches, just be consistent.

.

Shoulder to Extended Toe: Measure from the top of the right shoulder blade to the end of the longest toe pad (that should always be the third toe counting from the thumb or dew claw. To be consistent, do all the measurements on the right side of the body.
(Note: The photos in this document of the brown animal are being done on a live Belgian Tervuren dog.)

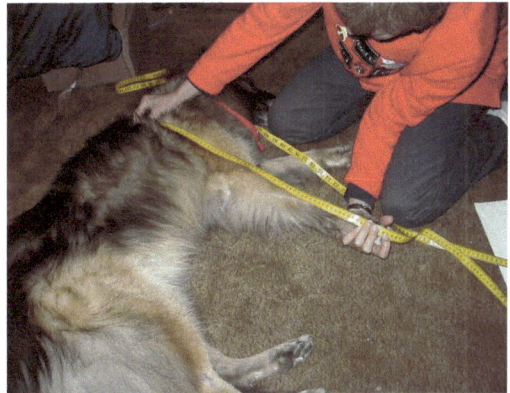

Shoulder to hip: Measure from the right shoulder blade right top of the pelvis.

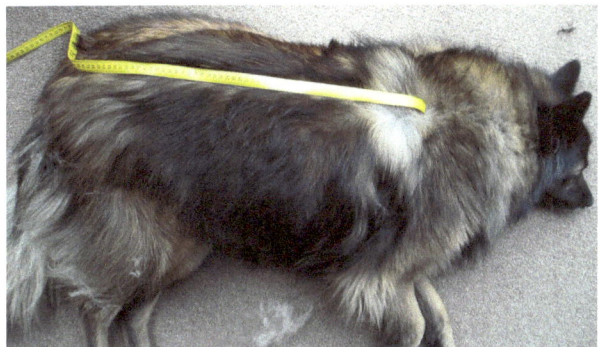

Head length: Measure from the tip of the nose to the bump in the skull that is the end of the saggital crest. This crest continues to grow throughout the animal's life so it will be larger on older animals.

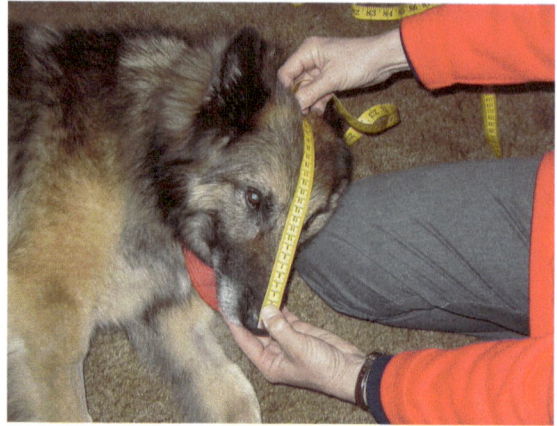

Nose Pad Height: Measure from the top surface of the nose to the end of the pointy triangle on the bottom of the nose

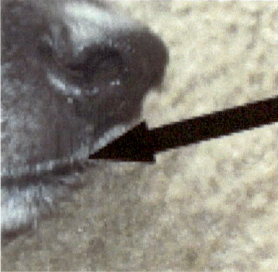

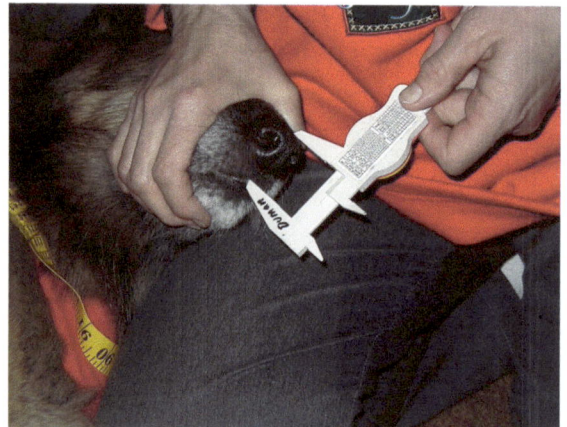

Nose Pad Width: measure across the widest part of the nose pad without squeezing the pad

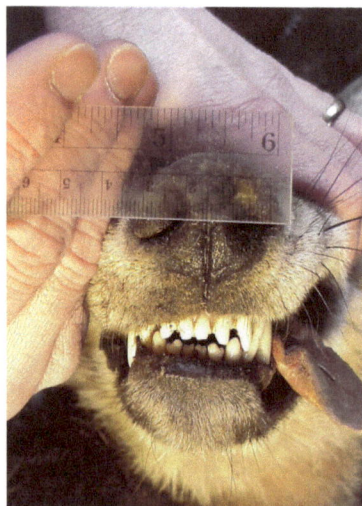

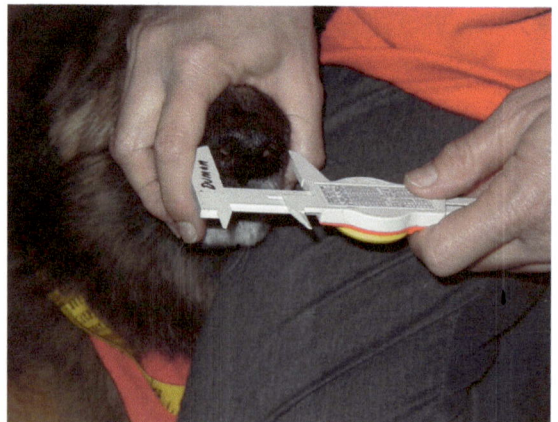

Upper Canine: Measure the width of the upper right canine at its base near the gums.

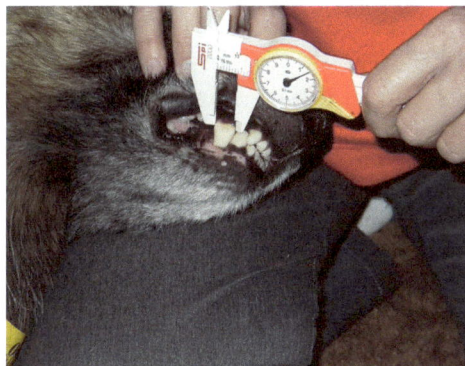

Canine Spread: Measure tip to tip of the upper canines

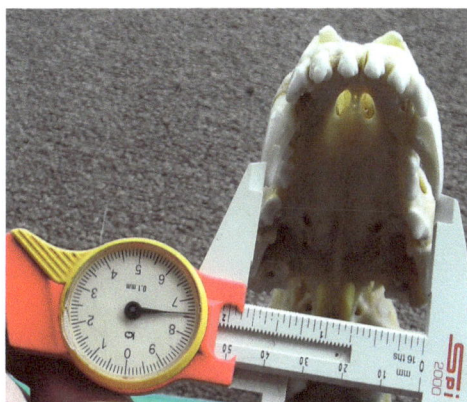

Ear Size: Measure from notch on the anterior of the ear to the tip of the ear not including the hair length

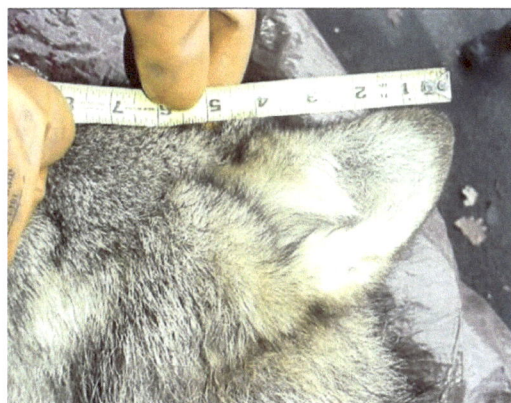

Foot Measurements: Use the right feet for measurement.

Front Pad Length: Measure from the most posterior of the foot pad to the end of the longest toe pad. (should be the third toe counting from the thumb (dew claw)

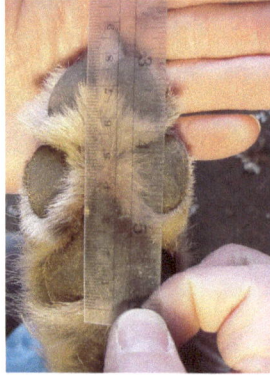

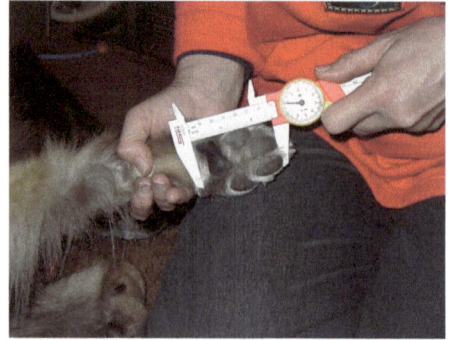

Front Pad Width: Measure the widest point across the toe pads with foot relaxed not squeezed

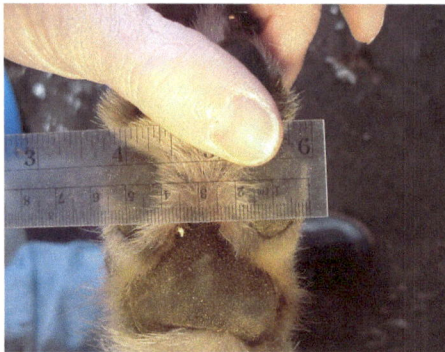

Rear Pad Length
Rear Pad Width: Measure the same as front feet

Rear Foot Length: Measure from the "heel" to the tip of the longest toe not including the toenail

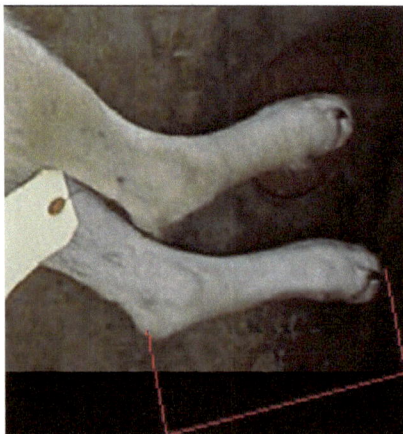

Great Lakes Area Wolves Compared to Coyotes

	Wolves	Coyotes
Weight	Rarely less than 50 pounds For adults.	Rarely more than 45 pounds
Height at Shoulder	70.cm – 86cm (28 – 34 inches)	50cm – 60 cm (20 - 24 inches)
Total Length (from nose to tip of tail)	Averaging 151 cm – 168 cm (59 to 66 inches)	Averaging 118 cm to 133 cm (46 to 52 inches)
General Appearance	Massive, extremely long legged	Delicate, medium sized, dog-like proportions with fox-like face
Color	Grizzled gray, black, faded to lighter shades, some tans, but Often darker in color	Grizzled gray usually with brown tones
Width of Nose Pad	At least 3.3 cm (1.5 inches)	No more than 2.5cm (I inch)
Diameter of Upper Canine Tooth at Base	At least 1.2 cm (Approximately 1/2 inch)	Less than 1.2 cm (Approximately 1/2 inch)
Muzzle	Heavier, long, but more Blocky in appearance	Long and pointed
Ears	Moderate sized, more round	Large in proportion to head size, and more pointed
Skull Length	Greater than 23.0cm (9.6 inches)	Less than 22.0 cm (8.7 inches)
Front Foot Length (length from rear of interdigital pad to foremost end of toe pad)	Usually greater than 8 cm (3.2 inches)	Less than 6.5 cm (2.6 inches)

Note: All data is for adult animals.

Great Lakes Area Wolves Compared to Coyotes

Figure 1. Captive wolf

Figure 2. Captive coyote

Compare the more fox-like face of the coyote with the less pointed heavier muzzle of the wolf. The coyote's ears are much larger in proportion to the size of its head.

Figure 3. Wild Michigan wolf in summer

Figure 4. Captive coyote in summer

In the summer, after shedding, the head of a wolf will appear large compared with the overall body size. The coyote's head will be more proportionate to the body size and the ears will appear very large.

Figure 5. Captive wolf

Figure 6. Wild coyote

Compared to a coyote, a wolf appears massive and very long legged.

www.ingramcontent.com/pod-product-compliance
Lightning Source LLC
Chambersburg PA
CBHW060818270326
41930CB00002B/77